Saturn

Saturn

Peter Murray

THE CHILD'S WORLD, INC.

Library of Congress Cataloging-in-Publication Data
Murray, Peter, 1952 Sept. 29–
Saturn/Peter Murray.
p.   cm.
Includes index.
Summary: Describes what is known about the second-largest planet in the
solar system and its rings and moons.
ISBN 1-56766-388-5 (alk. paper)
1. Saturn (Planet)—Juvenile literature. [1. Saturn (Planet)]
I. Title.
QB671.M87   1997
523.46—dc2   96-46674
CIP
AC

Photo Credits

COMSTOCK/COMSTOCK Inc.: 19
COMSTOCK/NASA: cover, 23
NASA: 10, 15,16, 29
Photri, Inc:: 2, 9, 13, 20, 24, 26, 30
Tony Stone Worldwide: 6

On the cover...

Front cover: Saturn is one of the most beautiful planets.
Page 2: This is what Saturn might look like from outer space.

Table of Contents

If you look at the sky on a clear night, you will see thousands of twinkling stars. If you look very carefully, you might find a bright spot of light that shines without twinkling. What you are seeing is a **planet** reflecting the light of the Sun. A planet is a world that circles around the Sun. If you know just where to look, you can find a yellowish spot of light that looks small and faint. This spot of light is the planet Saturn. It is more than 700 million miles from Earth!

There are many things to see in the night sky.

Saturn's Size

Nine planets circle around our Sun, and Saturn is the second largest. Only Jupiter is larger. More than 800 Earths could fit inside Saturn. It would be like filling a basketball with cherries!

Like all the other planets, Saturn spins around, or **rotates**, like a top. Even though Saturn is big, it rotates very fast. A day is the length of time it takes a planet to rotate once. Because Saturn spins so fast, a day on Saturn lasts only ten and one-half hours.

It is easy to see that Saturn is a huge planet.

Saturn has a blanket of stormy clouds thousands of miles thick. These clouds make up Saturn's **atmosphere**. The stripes and swirls you see on Saturn are the tops of the clouds. Saturn's cloudy atmosphere is always moving—sometimes with winds of over 1,000 miles per hour!

The stripes and swirls that cover Saturn are really clouds.

The Surface of Saturn

Saturn does not have a solid, rocky surface like Earth's. If you tried to land a spaceship on Saturn, you would simply sink through layer after layer of misty, icy clouds. The clouds would grow thicker and warmer as you got closer to the center of the planet. This center is called Saturn's **core**. It is hot and rocky. To reach it, you would have to go down through over 60,000 miles of swirling gases!

This picture shows the cloudy surface of Saturn.

Saturn's Moons

Out of all nine planets, Saturn has the most moons—at least 22! The biggest is *Titan* (TY–tun), which is 3,200 miles across. Titan is almost three times larger than Earth's moon. Most of Saturn's moons are made of rock and ice. *Rhea* (REE–uh), the second largest moon, is a huge ball of ice. Another moon called *Iapetus* (eye–AP–uh–tus) has a surface that is divided into two halves. One half is light and the other half is dark. The light area is probably ice, and the dark area is rock.

Iapetus has an icy side and a rocky side.

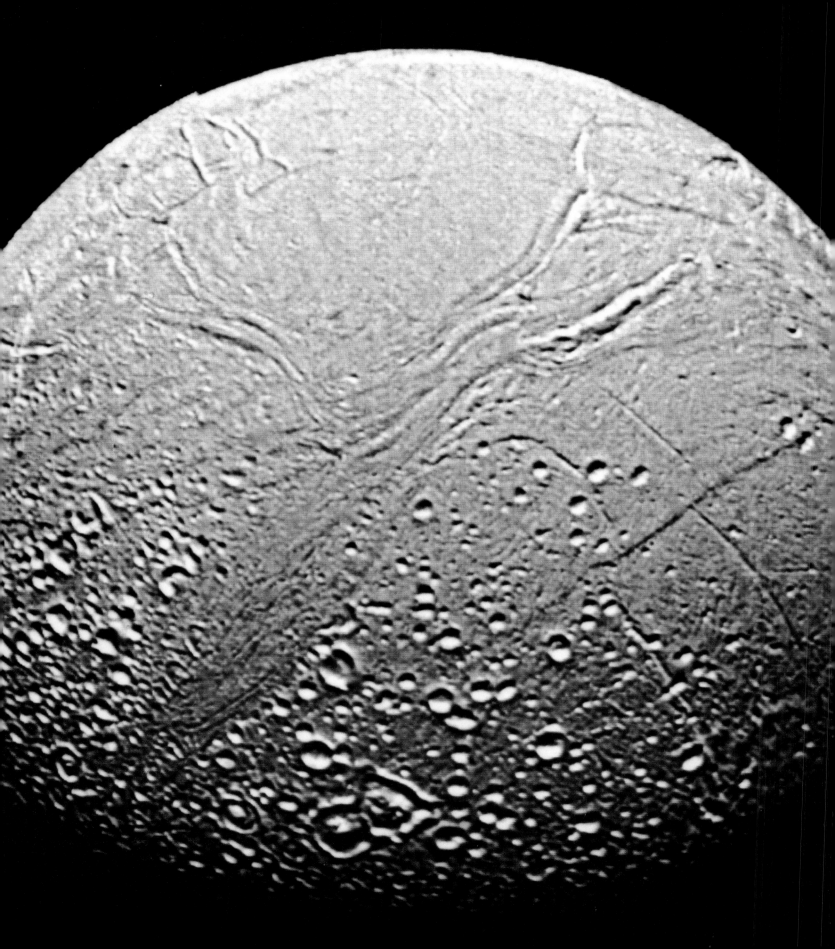

Dione (dy–OH–nee) is another one of Saturn's moons. It is a little smaller than the others. Dione is covered with rocks and large patches of ice. *Enceladus* (en–SELL–uh–dus) is also a small moon, but its smooth surface makes it the brightest. *Mimas* (MY–mus) is the closest moon to Saturn. It is only 200 miles wide, but it has a huge crater measuring 80 miles across.

This picture shows how bright *Enceladus* can be.

Saturn's other moons are oddly shaped rocks. They might be pieces of larger moons that broke apart. *Hyperion* (hy–PEER–ee–un) is the largest of these odd moons. It is a potato-shaped rock about 250 miles across. *Phoebe* (FEE–bee) is a faraway moon. It orbits the planet in the opposite direction from the other moons. Phoebe might have started out as a floating space rock called a **meteoroid**. Perhaps the meteoroid got caught by the force that pulls things towards planets. This force is called **gravity**.

Saturn's moons probably got caught by the planet's gravity .

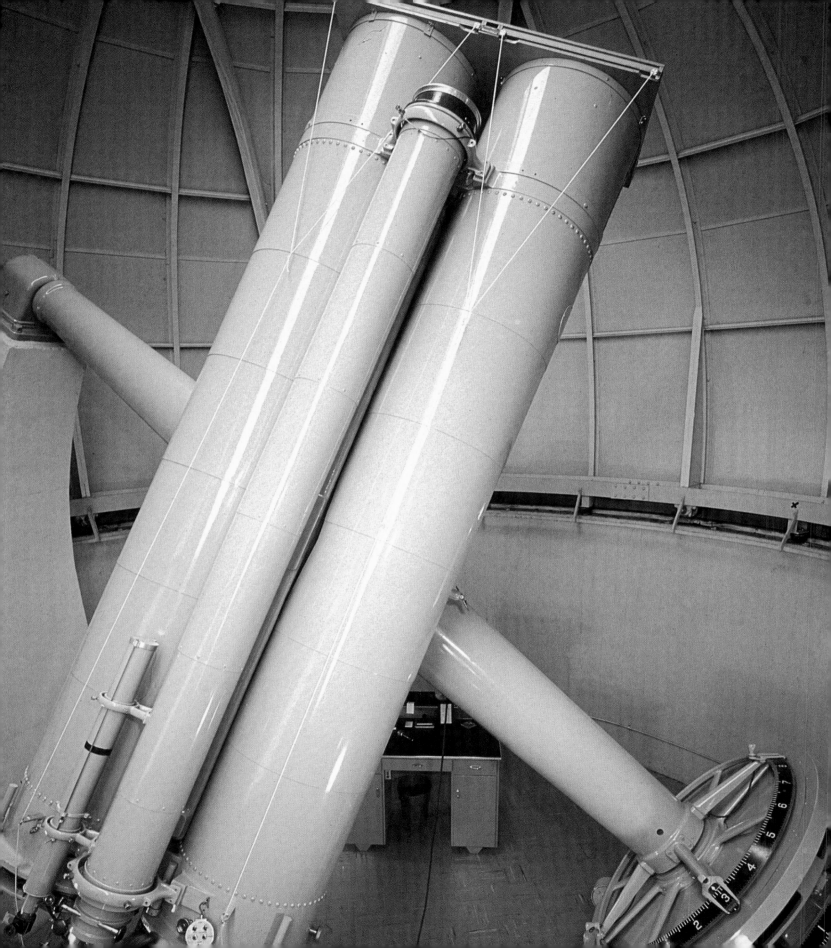

Astronomers

People who study the planets and stars are called **astronomers**. They use special tools, called **telescopes**, to see objects that are very far away. Telescopes make these objects look bigger and closer.

A long time ago, an astronomer named Galileo (ga–luh–LAY–oh) was looking through his telescope. He noticed something that made Saturn different from all the other planets. He saw two strange lumps, one on each side of Saturn!

Galileo's telescope was not good enough to show him what the lumps really were. A few years later, a man named Cassini (cuh–SEE–nee) looked at Saturn through a better telescope. He discovered that these strange lumps were actually many rings that went around the planet.

From far away, Saturn's rings look like huge lumps on the planet.

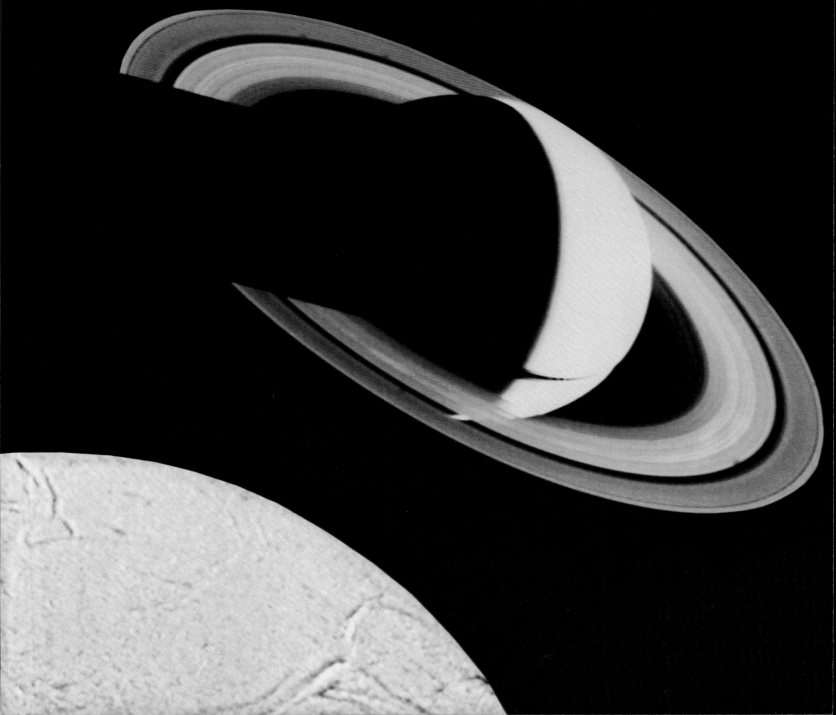

Saturn's rings circle around the planet like hula hoops. The rings are not solid. Instead, they are thin layers made up of billions of small pieces, or **particles**, of dust, rocks, and ice crystals. Some of the rings are made up of fine dust. Others contain rocks as big as houses!

Saturn's rings circle the planet like hula hoops.

Saturn's rings are over half a million miles across, but less than a mile thick. In fact, you can see right through them! There are hundreds of thin rings. Each goes around the planet in a circle, called an **orbit**. The inside rings orbit Saturn the fastest—once every six hours. The outside rings move more slowly. They take about 14 hours to circle the planet.

This picture shows how thin Saturn's rings are.

A Mystery

How did Saturn's rings form? Some scientists believe the rings are the crushed pieces of a moon. Perhaps the moon moved too close to Saturn and was pulled apart by gravity. Other scientists believe the rings are left over from when the planet formed. The true beginnings of Saturn's rings remain one of the biggest mysteries of outer space.

Many people have different ideas about how Saturn's rings formed.

One day men and women may travel to the other planets. Perhaps someday, you or your children will stand on the edge of Mimas's giant crater. High above you, the striped and swirling clouds of Saturn will fill most of the sky. Perhaps you will go there to discover how the rings were formed. Maybe you will want to explore the surface of the strange moon, Titan. Or maybe you will just want to learn what lies beneath Saturn's clouded surface. There is still much to learn about this wonderful and puzzling planet!

Saturn's colors and rings make it a very beautiful planet.

Glossary

astronomers (uh–STRON–uh–mers)
Astronomers are people who study the planets and the stars. Astronomers want to learn more about Saturn.

atmosphere (AT–mus–feer)
An atmosphere is a layer of gases that surrounds a planet. Saturn's stormy atmosphere is always moving.

core (KOR)
The core is the center of a planet. Saturn's core is very hot and rocky.

gravity (GRAV–ih–tee)
Gravity is the force that pulls things toward a planet or a moon. Saturn's moon Phoebe might have been caught by Saturn's gravity.

meteoroid (MEE–tee–uh–royd)
A meteoroid is a rock that floats around in outer space. The moon Phoebe might have been a meteoroid at one time.

orbit (OR–bit)
When something circles all the way around something else, it makes an orbit. Saturn's rings orbit around the planet.

particles (PAR–ti–kuls)
Small pieces of things are called particles. Saturn's rings are made of particles of dust, rock, and ice.

planet (PLA–net)
A planet is a world that circles around a star. Saturn is one of the nine planets that circle our Sun.

rotate (ROH-tate)
When something rotates, it spins all the way around, like a top. Saturn rotates once every ten and one half hours.

telescopes (TEL–uh–skopes)
A telescope is a special tool that helps people see things that are very far away. Telescopes helped early scientists learn more about Saturn's rings.

Index